Let's Make

Slime

by Katie Chanez

NORWOOD HOUSE PRESS

Norwood House Press

For information regarding Norwood House Press, please visit our website at:
www.norwoodhousepress.com or call 866-565-2900.

Hardcover ISBN: 978-1-68450-839-6
Paperback ISBN: 978-1-68404-623-2

LIBRARY OF CONGRESS CATALOGING-IN-PUBLICATION DATA

Names: Chanez, Katie, author.
Title: Let's make slime / Katie Chanez.
Description: Chicago : Norwood House Press, 2021. | Series: Make your own: science experiment! | Includes index. | Audience:
    Grades 2-3.
Identifiers: LCCN 2019053627 (print) | LCCN 2019053628 (ebook) | ISBN 9781684508396 (hardcover) | ISBN 9781684046232
    (paperback) | ISBN 9781684046294 (ebook)
Subjects: LCSH: Science--Experiments--Juvenile literature.
Classification: LCC Q164 .C418 2020 (print) | LCC Q164 (ebook) | DDC 507.8--dc23
LC record available at https://lccn.loc.gov/2019053627
LC ebook record available at https://lccn.loc.gov/2019053628

328N—072020
Manufactured in the United States of America in North Mankato, Minnesota.

Contents

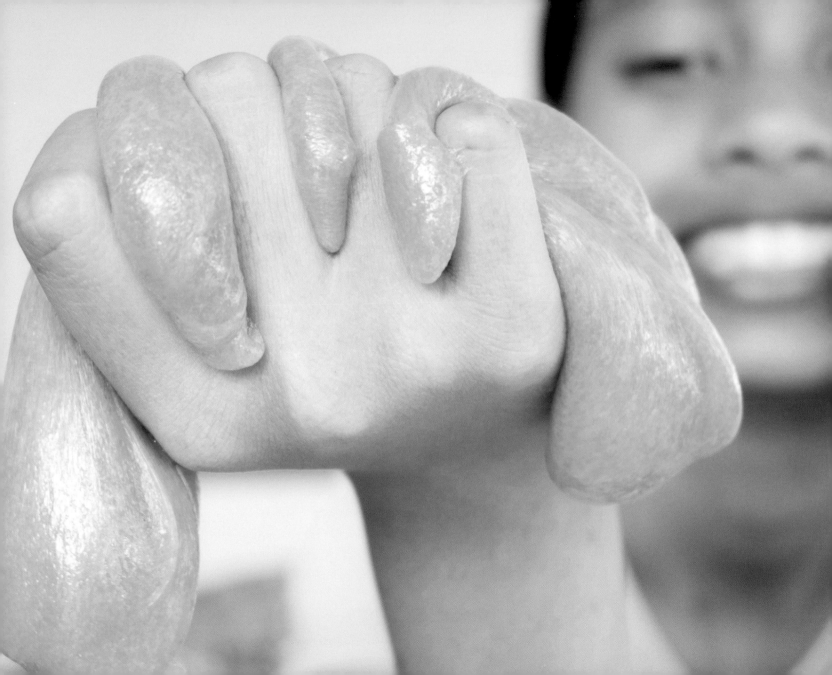

Slime is fun to squeeze.

CHAPTER 1

All about Slime

Slime is a strange liquid. It can be rubbery and slippery. It can be sticky and gooey. It stretches and bounces. It is fun to play with. Slime has been sold as a toy since 1950. It comes in many colors. Some slime has glitter or beads to change how it feels. People of all ages enjoy playing with slime.

Slime is not just a toy. It can be found in nature too! Many kinds of animals make slime. It helps keep them safe and healthy. Snails and slugs create sticky slime. It helps them climb to reach food and shelter. Catfish make slime. It helps them heal faster when they are hurt. Hagfish make slime when a predator tries to eat them. This slime chokes the predator. Then the hagfish can escape.

There is also a type of living thing called slime mold. Slime molds are not plants or animals. Instead, they are a type of living slime. Slime mold grows in moist areas such as forests. Scientists have discovered over 900 types of slime molds.

Even the human body makes a type of slime! This slime is called **mucus**. Mucus traps germs that can make a person sick. The body

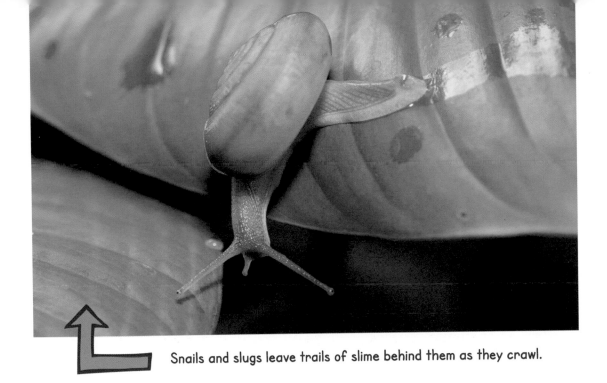

Snails and slugs leave trails of slime behind them as they crawl.

creates more slimy mucus when a person gets sick. This extra mucus can lead to a runny nose or a cough.

All objects are made of matter, including slime. All matter is made of tiny **atoms**. Atoms connect to form **molecules**.

Matter commonly forms in three states: solids, liquids, and gases. Solids have tightly packed atoms. The molecules do not move. This gives the solid a fixed shape. Liquids do not have tightly packed atoms. The molecules can move around and slide past each other. The liquid changes its shape based on the shape of the container it is in. Slime usually acts like a liquid. It flows. It changes its shape. But sometimes it acts like a solid.

Liquids have different levels of **viscosity**. Water has a low viscosity. It flows easily. Molasses has a high viscosity. It does not flow easily. Adding or taking away heat changes the viscosity for

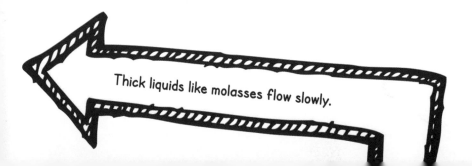
Thick liquids like molasses flow slowly.

most liquids. Pancake syrup can be hard to pour when it is cold. Heating the syrup causes it to flow faster.

But not all liquids follow these rules. Slime is one of these rule-breakers. Some liquids change their viscosity when **stress** is applied. These liquids act like a solid in certain conditions. Some liquids act solid until stress is added. For example, ketchup easily gets stuck in a bottle. Shaking the bottle adds stress. Then the ketchup flows more easily. Other liquids do the opposite. They act like a solid only when stress is added. Quicksand is a mix of water and fine sand. Trying to quickly pull an object out of it adds stress. The quicksand thickens. Gently pulling a little bit at a time is the only way to get an object out. Slime is another liquid that can act like a solid when stress

Slime acts like a liquid or solid based on how you play with it.

is added. It feels solid if poked quickly. It feels liquid if touched gently. Slime can bounce when it is dropped. But it will ooze if left alone on a table.

Slime gets its rubbery texture from **polymers**. Polymers are long chains of molecules. One molecule connects to others. The chains can be made of the same type of molecule. Sometimes they are made of different kinds. Many items found in nature are polymers, including rubber and fingernails. People also make polymers. Plastic is one of the most common types. The polymers in your slime come from glue. Because the glue is a liquid, these polymers slide past each other. This lets the glue flow out of the bottle. Slime gets its gooiness from a **chemical reaction**. A chemical reaction occurs when two or more molecules interact. Your reaction will happen when you add a cleaning product called borax to the glue. The borax connects the polymers. They can no longer move easily. Rubbery slime forms!

Glue and Borax
Chemical Reaction

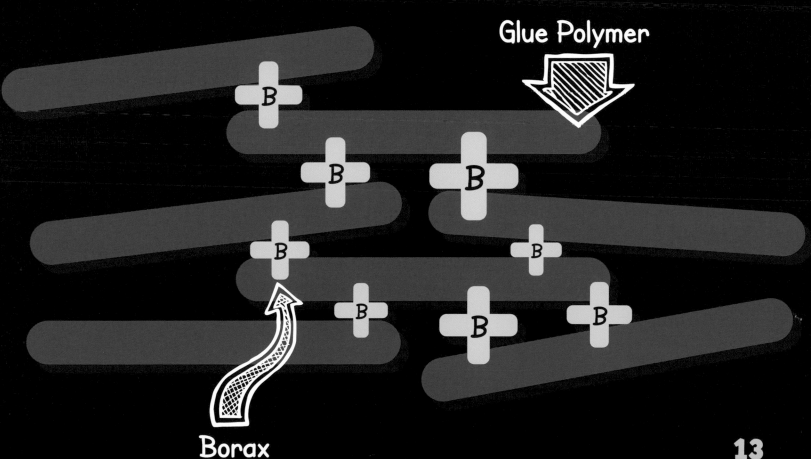

Glue Polymer

Borax

CHAPTER 2

Make Your Own Slime

Homemade slime needs a chemical reaction to form. For this reaction to happen, you need the correct ingredients. You need ingredients that will get stretchy and rubbery when combined.

The first ingredient you need is school glue. School glue is made of long polymers. These molecules are strong and flexible. This will be the base of your slime. The polymers will make the slime stretchy. But the glue is very thick. It is hard to mix in other ingredients. Adding water creates a glue **solution**. The solution is thinner than plain glue. Now it is easier to add the other ingredients.

Another key ingredient in slime is borax. Borax is a white powder that forms naturally. It has many uses. It is commonly used to clean. Its main ingredient is boron. Boron is often found in rocks. The borax dissolves in water. This forms another solution. Combining this mixture with your glue solution will cause the chemical reaction.

A **chemical bond** forms when the borax is added to the glue. The borax begins to connect the polymers in the glue. This process

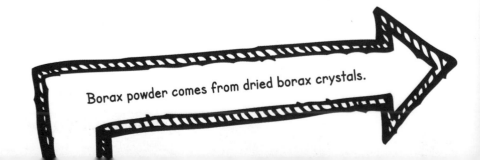

Borax powder comes from dried borax crystals.

is called **cross linking**. The molecules hook together like tangled noodles. They can no longer flow freely. The slime gets rubbery as more polymers connect. The water in the slime helps the polymers slide after they have been linked. This is why the slime oozes. Your slime would harden without the added water.

Make your slime away from carpet or fabric. The sticky slime is hard to clean off soft surfaces. Use a clean workspace to keep dirt or crumbs out of your slime. It is important to wash your hands when you are done. This gets any remaining slime or color off your hands. You can keep your slime when you are done. Store it in an airtight container for up to two weeks in the refrigerator.

Materials Checklist

- ✓ 2 mixing bowls
- ✓ 1 bottle of school glue
- ✓ Water
- ✓ Borax

- ✓ Food coloring (optional)
- ✓ 1/2 cup measuring cup
- ✓ 1/4 teaspoon measuring spoon
- ✓ Spoon or craft stick for stirring

You can use water that has been heated in a microwave or use hot water from a sink.

CHAPTER **3**

Science Experiment!

Now that you know what makes slime act the way it does, put that knowledge to use and make your own!

1. Measure 1/2 cup of warm water. Pour it into one of the bowls.

2. Measure 1/4 teaspoon of borax. Mix it into the water with the spoon or craft stick until the borax has dissolved. You can also mix in a few drops of food coloring if you want your slime to have color.

3. Measure 1/2 cup of glue and pour it into the empty bowl.

4. Add 1/2 cup of water to the glue and stir until they are completely mixed.

5. Carefully pour the glue solution into the bowl with the borax solution.

26

6. Stir everything together until it becomes one mixture. Once it is hard to stir, take your slime out of the bowl. It is okay if there is extra water left in the bowl. Use your hands to **knead** the slime. Continue to knead until your slime becomes smooth.

7. Now it is time to play with your slime! See how it moves and stretches. Pull the slime quickly. Now pull it slowly. See how the slime reacts.

Make It Better!

Congratulations! You have made slime. Now see if there are ways to improve it. Use any of these changes and see how they improve your slime.

- The recipe lists a specific amount of borax to use. Does changing the amount of borax change the slime's texture?

- Boron can be found in many items besides borax. Two examples are contact lens solution and liquid starch for ironing. How do you think these will change your slime?

- Mix in other items after making your slime. You can use glitter, beads, marbles, or similar materials. How does this change your slime?

Can you think of any ways that you could improve or change your slime to make it better?

Glossary

atoms (AT-umz): Tiny things that are building blocks of all matter.

chemical bond (KEM-uh-kuhl BOND): A force that connects atoms or molecules together.

chemical reaction (KEM-uh-kuhl ree-AK-shuhn): A process in which the atoms in ingredients rearrange into something else.

cross linking (KROSS LINK-ing): When an atom or molecule connects to another atom or molecule.

knead (NEED): To mix ingredients together by pressing or squeezing with one's hands.

molecules (MOL-uh-kyools): Groups of connected atoms.

mucus (MEW-cuss): A slimy goo that animals and humans make to protect parts of their bodies from germs.

polymers (PAH-luh-murz): Long, repeating chains of molecules.

solution (suh-LOO-shun): A mixture of a solid, liquid, or gas combined with another solid, liquid, or gas.

stress (STRESS): Force added to an object.

viscosity (vis-COS-it-ee): How quickly or slowly a liquid flows.

For More Information

Books

Ellen Lawrence. *Creeping Slime: Slime Molds.* New York, NY: Bearport Publishing, 2019. In this book, students can learn about slime molds.

Julia Garstecki. *Science in a Jar: 35+ Experiments in Biology, Chemistry, Weather, the Environment, and More!* Beverly, MA: Quarry Books, 2019. This book explores physical and life sciences using many different experiments.

Natalie Rompella. *Experiments in Material and Matter with Toys and Everyday Stuff.* North Mankato, MN: Capstone Press, 2016. Students can try different experiments using household objects to learn about matter and its properties.

Websites

DK Find Out: Solids, Liquids, and Gases (https://www.dkfindout.com/us/science/solids-liquids-and-gases/) Students can learn about matter and its different states.

Imagination Station: Oobleck (https://www.imaginationstationtoledo.org/educator/activities/oobleck) Students can learn to make another substance that has the properties of both solids and liquids.

PBS Learning Media: States of Matter (https://tpt.pbslearningmedia.org/resource/idptv11.sci.phys.matter.d4kmat/states-of-matter/) This video teaches students about matter and how it can change forms.

Index

About the Author

Katie Chanez is a children's book writer and editor originally from Iowa. She enjoys writing fiction, playing with her cat, and petting friendly dogs. Katie now lives and works in Minnesota.